Sylvester David Willard

Conservative Surgery

With a list of the medical and surgical force of New York in the War of the

Rebellion, 1861-2 : to which is added a brief notice of the hospitals at

Fortress Monroe and White House, Virginia

Sylvester David Willard

Conservative Surgery
With a list of the medical and surgical force of New York in the War of the Rebellion, 1861-2 : to which is added a brief notice of the hospitals at Fortress Monroe and White House, Virginia

ISBN/EAN: 9783337410476

Printed in Europe, USA, Canada, Australia, Japan

Cover: Foto ©berggeist007 / pixelio.de

More available books at **www.hansebooks.com**

CONSERVATIVE SURGERY,

WITH A LIST OF THE

MEDICAL AND SURGICAL FORCE OF NEW YORK

IN THE WAR OF THE REBELLION,

TO WHICH IS ADDED A BRIEF NOTICE OF THE HOSPITALS AT
FORTRESS MONROE AND WHITE HOUSE, VIRGINIA.

By SYLVESTER D. WILLARD, M. D., OF ALBANY,

SECRETARY OF THE MEDICAL SOCIETY OF THE STATE OF NEW YORK,
HONORARY MEMBER OF THE CONNECTICUT MEDICAL SOCIETY, &c.

ALBANY:
CHARLES VAN BENTHUYSEN, PRINTER.
1862.

REPRINT FROM TRANSACTIONS OF MEDICAL SOCIETY OF THE STATE OF NEW YORK, 1862. READ BEFORE THE ALBANY INSTITUTE, APRIL 29, 1862.

CONSERVATIVE SURGERY:

LIST OF THE MEDICAL AND SURGICAL FORCE OF NEW YORK,
IN THE WAR OF THE REBELLION, 1861, '62. BRIEF NOTICE
OF THE HOSPITALS AT FORTRESS MONROE AND WHITE-
HOUSE, VIRGINIA, ETC.

As the wealth of a State consists in the number of its popula-
tion, so the strength of an army depends not only upon its num-
bers, but in their ability to put in exercise the munitions of war-
fare. And this ability depends not only upon the skill of com-
manding officers to direct the movement of troops, but in the
physical strength of troops to endure wearisome marches and
fierce engagements in battle. The first and most important
qualification of soldiers then, is health ; and to exercise a surveillance over them and to keep them in health, is the duty of
the surgeon. In the formation of an army the surgeon is as important and indispensable as the general.

Nor do we any longer associate with the army surgeon the
idea of a man who sees before him hundreds of wounded soldiers,
with no other feeling than the variety and opportunity they
afford him for an almost indiscriminate use of the catling or scal-
pel, and who only sees in mutilated limbs and broken bones what
can be amputated, instead of what possibly *may* be saved ; who
seeks to dispatch a case suddenly, by giving it a positive character, rather than by patient waiting to aid nature in its restora-
tion. That the army is a place to learn surgery, has been an
opinion very generally received ; that it has been learned at the
expense of many a poor soldier, whose limbs might have been
saved, may as truthfully be added.

The events which within the last year have plunged our nation
into the evils of a civil war, are calculated to develop a new era
in the history of military surgery, and install the military surgeon with new dignities and more humane duties. His chief
duties are no longer those of the mechanical or skilful operator,
but they are extended to the protection and almost parental care
of the soldiery.

The principles of hygiene and sanitary science are now un-

4

folded, and thrown as a protection around the army to insure it against the diseases of the camp, which have ever proved more fatal in the march of armies than the bullets and shells of the enemy. We need only to cite the instances of sickness and mortality in the French army at the Crimea, as reported by Dr. G. Scrive, the Surgeon General of the army. That war lasted for twenty-four months, summer and winter, without any intermission. The entire number of French troops that were sent to the East amounted to 309,268 men. Of this number, 200,000 were received in ambulances and hospitals, for medical.aid, 50,000 for wounds, and 150,000 for diseases of various kinds. The total mortality was 69,229; or 22½ per cent. 16,325 of these died of wounds, and about 53,000 from diseases. Thus, those by diseases, it will be seen, are more than three times the number of those destroyed by wounds. There were of ordinary wounds, 2,185 ; gunshot wounds, 22,891 ; frost-bitten, 3,472 ; typhus fever, 3,840; cholera, 8,196 ; scurvy, 17,576 ; feverish, 63,124 ; venereal, 241 ; itch, 124. There was a great mortality from scurvy and frost bites. Of the 25,076 cases of ordinary and gunshot wounds, it will be observed that there was the immense mortality of about seventy-five per cent., notwithstanding all that skillful surgery could do to avert such melancholy results. Many of these cases terminated doubtless as the sequels of operations, by traumatic fever, hemorrhage, excessive suppuration, spasms, erysipelas, gangrene, pyemia and tetanus. Pyemia at the Crimea was one of the chief sources of danger after operations. How many of that 16,325 might have been saved by a more conservative surgery, and thus avoiding the sequelæ of operations as dangerous as the wounds themselves, is a problem perhaps not easily calculated. But might not an increased vigilance in the sanitary and hygienic regulations have averted the pneumonia, and the typhus and the fevers, by which the fearful number of 150,000 troops were prostrated in a period of two years, and of which 53,000 died—a mortality of 33½ per cent; and doubtless the lives of many others were shortened by exposure and disease that developed with fatal results after the termination of the war. This comparison is not instituted for the purpose of deprecating skillful surgery. We award to it the relief it brings, and the grand results it achieves.

In the hands of the conservative surgeon neither the officer or those of his command have anything to fear. In the hands of the bold, dashing operator, and such sometimes find their way to the surgical staff, there is more cause for apprehension. But the point intended to be illustrated by the comparison is, that if all cases requiring capital operations, doubtful in their results, were abandoned to nature, and the skill of the surgeon directed to the prevention of disease, by discovering and avoiding the sources of malaria, by careful inspection of the dress and diet, the amount of fatigue incurred, and the general hygienic condition of the camp and hospital, whereby the sickness of the 150,000 might be diminished and the fearful mortality of $33\frac{1}{3}$ per cent. lessened, the strength of the army would be increased and therewith a saving of human life greater than that hoped for by the most brilliant surgical operations. This position in favor of conservative surgery is elucidated and sustained by the remarks of a surgeon in the volunteer service of great practical experience, who writes from a distant field of service, where he has had opportunity to put into practice the precepts* of military surgical authorities : " I conjure you," says he, " in the august name of that humanity which should be identified with the title of surgeon, to guard young surgeons against yielding to their eagerness to *cut*, and also to counsel against primary amputations, particularly in the thigh where there is a large wound of the soft parts combined with fracture of the bone. Of the three 'primary' amputations performed in my presence yesterday, the subjects all died, one of them in the very act of the operation. In each of these cases, with the rules of army surgery stated by the eminent army operators fresh in my mind, I silently dissented from the practice." Should the dissent have been silent ? Could not the unfavorable prognosis have been foreseen and the operations avoided or delayed until they gave some slight promise of success ?

The call for surgeons to supply the volunteer troops of the State of New York, has been responded to with a zeal worthy of a noble profession, with an enlightened and liberal patriotism that sheds a luster upon the escutcheon of the Empire State, and which has already reflected honor upon our nation, and given distinction by its heroism and fidelity to our national arms. Among the number are some of the most accomplished surgeons in the country, and one† who is among the distinguished members of our society, to whom we point with just pride.

* Am. Med. Times, March 29, 1862, p. 182.   † Prof. Frank H. Hamilton.

It has been my object to collect from official sources the entire medical and surgical force from the State of New York, which has been engaged in this unhappy war, and to arrange it in such a form as to give it convenience for future historical reference to the whole profession and to the State. No place could be more appropriate for its publication than the Transactions of our Society. My first thought was to refer, by foot notes, to the many interesting articles published in the journals of the day, by the various members of the medical staff, but I soon found that such a plan would be necessarily incomplete on account of the limited number of medical periodicals that might come under my own eye, and it was accordingly reluctantly abandoned.

No ordinary interest would now be attached to a full account of the surgeons of the Revolution and a history of their operations, or to those which our State furnished in the war of 1812. Whatever either accomplished, their names and labors find no record on the page of history.

In the number who compose the medical staff of the State of New York, I do not include those citizens of this State who were already in the regular army or navy service when the war began, or those engaged in the three month militia service, but

1st. Those who were examined by the Naval Board after the commencement of the war.

2d. Those who were examined by the Army Board after the commencement of the war.

3d. The Sanitary Commission.

4th. Those examined by the Medical Board at Washington.

5th. Those examined by the Medical Board at Albany.

The sanitary commission, as is well known from their frequent published documents, was organized at Washington on the 13th of June, 1861, in accordance with a suggestion made to Mr. Cameron, the Secretary of War, by the acting Surgeon General, R. C. Wood, on the 22d of May previous. The suggestion met the approval of the Secretary of War and received the approbation of the President, Mr. Lincoln, on the 13th of June, the same day on which the commission was organized. Its object was to lessen the pressure upon the medical bureau in view of so large an army, and to direct the "intelligent mind of the country to practical results connected with the comforts of the soldier by preventive and sanitary means." It includes in its range all that refers to the health, comfort and morale of the troops; by seeing how far a volunteer force may be speedily brought to conform to the standard regulations of the regular army; by making scientific inquiry into all that relates to camp grounds, clothing, tents, cooks, cooking and diet; by exercising precaution against excess of heat and cold; by guarding against the influences of malaria and infection; by providing early and ample comforts for the sick and wounded, by general attention to military hospitals; these investigations to be guided by the highest medical and military experience and foresight.

The commission have labored diligently and achieved much towards carrying out their original designs. They have appointed a large number of associate members* in all the loyal states to co-operate with them in their benevolent and patriotic measures; and already more than forty documents, some of them written with marked ability, have been published under their authority.

The record of brigade surgeons, which was obtained from the Surgeon General's department in Washington, shows that of the twenty-two who were examined by the medical board in Washington, six had already been examined in Albany, and appointed to the charge of regiments.

The examining board at Albany, was organized on the 19th April, 1861, by His Excellency, Governor Morgan, upon the suggestion of Surgeon General Vanderpoel. The board consisted of Drs. Alden March, Thomas Hun, and Mason F. Cogswell, gentlemen well known for their high toned professional accomplishments, their sterling integrity, and their earnest patriotism. Up to the 10th December, the board had examined 468 applicants, 228 of whom were accepted as qualified to act as surgeons, and 137 as assistant surgeons.

The plan of the examination was by a series of printed questions that were placed before the applicant, answers to which he was required to write within a given time, as concisely and completely as possible, without consulting books or persons. The topics embrace anatomy, surgery, chemistry, theory and practice of medicine, and therapeutics.

The answers were carefully examined together with testimonials of character and skill, and in view of the same the grade of the applicant was determined. This board remark in its report to the Surgeon General, that "It is believed the list embraces a

* List of members; Sanitary Document No. 34; Dec. 7, 1861.

body of men possessing that character, education, practical skill and experience which all so earnestly desire, may be secured in behalf of the health and the lives of our volunteer forces."

The inquiry has been so often repeated as to the character of these examinations that I make no apology for appending a copy of the series of questions. They may be found also in the report of Surgeon General Vanderpoel, to the Governor and Commander-in-chief of our State forces, and are as follows :*

Copy of questions submitted Applicants by Examining Board.

First series.

Each candidate will, without reference to books, furnish written answers to as many of the following questions as the alloted time will allow. The answers should be concise, and at the same time as complete as possible.

Each answer should be numbered to correspond with the number of the question.

The paper containing the answers is to be signed, and, together with this sheet, inclosed in a sealed envelope, on the back of which the name and address of the candidate are also to be written.

1. Describe the course and relative position of the femoral artery ; also the operation of ligature of the femoral artery.

2. Answer the same questions in regard to the brachial artery.

3. Give the names, situation and distribution of the principal nerves of the upper extremity.

4. Describe the operation of amputation of the thigh. Describe the operation of amputation of the leg. Describe the operation of amputation of the forearm.

5. Describe the dressing and subsequent treatment, and the accidents which may follow these operations.

6. Give the diagnostic signs of compression and concussion of the brain, and the general treatment applicable to each.

7. Describe the accidents accompanying incised wounds, and the treatment.

8. Describe the characters and treatment of lacerated wounds.

9. Give the characters and accidents peculiar to gunshot wounds, and the general treatment.

10. What are the rules for amputation in cases of gangrene ?

11. Under what circumstances is traumatic erysipelas liable to come on, and how it is to be prevented and treated ?

12. What are the symptoms of shock or collapse, following severe injuries ? Give the treatment of this condition.

13. Give the hygienic and medical treatment of dysentery occurring in camp life.

* Assembly Doc. No. 12, 1862.

14. What are the constitutional disturbances caused by burns? Give the general and local treatment.

15. Give the chemical composition, medical uses and mode of administration of the following substances ; calomel, corrosive sublimate, iodide of potassium, epsom salts, sulphate of copper, lunar caustic.

Copy of questions, second series.

1. Describe the different dislocations of the os humeri—the diagnostic signs of each, and the mode of reduction in each case.

2. Answer the same questions in regard to the dislocations of the hip joint.

3. Describe the course and relative position of the arteries of the forearm and hand.

4. Describe the operation of ligature of the anterior tibial artery. Describe the operation of ligature of the radial artery. Describe the operation of ligature of the external iliac.

5. Describe the operation for strangulated inguinal hernia, dressing and subsequent treatment.

6. Give the most important means of arresting hemorrhage from incised wounds; also from punctured wounds.

7. Give diagnostic signs and treatment of fracture of the lower portion of the radius ; also of the lower portion of the fibula.

8. Give the symptoms of scurvy, its causes, mode of prevention and treatment.

9. Give the symptoms and physical signs of the different stages of pneumonia.

10. Describe the danger of penetrating wounds of the thorax; the symptoms of wounds of the lung ; and the general management of such accidents.

11. Give the general character and treatment of gunshot wounds.

12. Give the treatment of wounds of the intestines.

13. What is meant by pyœmia ? Under what circumstances does it occur ? How is it to be recognized and treated ?

14. Give the medicinal properties, modes of administration and doses of the following substances : aloes, jalap, calomel, opium, tartarized antimony, sulphate of zinc.

Copy of questions, third series.

1. Describe the symptoms and course of typhoid fever, its anatomical lesions and treatment.

2. Give the causes, symptoms and treatment of bilious remittent fever ; also of intermittent fever.

2

3. Give the symptoms, physical signs and anatomical lesions of pericarditis.

4. Give the symptoms of hectic fever, the circumstances under which it occurs, and its treatment.

5. Describe phlebitis, its causes and its consequences.

6. Describe the causes and consequences of varicose veins of the leg, and the mode of management.

7. Describe some of the principal acute inflammations of the eye; explain the tissues involved, and give the treatment.

8. Describe the primary, secondary and tertiary forms of syphilis, the diagnosis and treatment.

9. Describe the dangers and general mode of treatment of fractures, simple, compound and comminuted.

10. Describe the apparatus necessary for dressing a fracture of the femur, and its mode of application.

11. Answer the same question in regard to fracture of the tibia.

12. Give the rules for applying ligatures to large arteries, and the subsequent treatment.

13. Describe the course and situation of the large arteries and veins of the neck.

14. Give the symptoms and physical signs of phthisis pulmonalis in its early stages.

15. Give the medicinal uses, the doses and mode of administration of the following substances: digitalis, extract of belladonna, nitrate of potash, tartar emetic, chloroform, cod liver oil.

Copy of questions, fourth series.

1. Describe the origin, position and distribution of the sciatic nerve.

2. Describe the origin, course and general distribution of the fifth pair of cranial nerves.

3. Describe the ligaments of the hip joint.

4. Describe the origin and course of the right and left primitive carotid arteries.

5. Enumerate and describe the membranes of the brain.

6. Under what circumstances is the operation of trepanning necessary? Describe the operation and subsequent treatment.

7. Describe the operation of ligature of the femoral artery, the subsequent treatment and the accidents which may follow the operation.

8. Give the causes and treatment of fistula in ano.

9. Give the diagnostic signs, prognosis and treatment of fracture of the femur within the capsular ligament.

10. Describe the operation of amputation of the thigh, the dressing and subsequent treatment.

11. Describe the accidents which may follow amputation.

12. Give the symptoms, course and anatomical lesions of typhoid fever.

13. Give the symptoms, physical signs and modes of termination of acute and chronic pleurisy.

14. Describe the organic lesions which give rise to dropsical effusions.

15. Give an account of the principal constituents of the blood.

16. Give the medicinal uses and modes of administration of the following substances : nitrate of silver, senna, sulphate of magnesia, colchicum, digitalis, opium, hyoscyamus, corrosive sublimate.

Copy of questions, fifth series.

1. Give the commencement, course, termination and relations of the jugular veins.

2. Give the diagnostic symptoms of paralysis of the facial nerve.

3. Describe the operation of excision of the knee joint, and also of the elbow joint, and state in what cases these operations should be preferred to amputation.

4. In cases which admit of a choice, which method is to be preferred in amputation of the foot? Describe the operation.

5. Give the diagnosis and treatment of fractures of the clavicle.

6. Give the differential diagnosis between hernia and varicocele, and describe the operation for the radical cure of the latter.

7. Give the causes and treatment of hemoptysis.

8. Give the causes and treatment of retention of urine, and describe the method of introducing the catheter.

9. Give the symptoms, cause and treatment of acute rheumatism and its complications.

10. Give the symptoms and treatment of diphtheria and its sequelæ.

11. Describe the primary, secondary and tertiary forms of syphilis, the diagnosis and treatment.

12. Give the tests for albumen, and also for sugar in the urine.

13. Write out in full a prescription for a purgative pill, a diuretic mixture, and a cough mixture.

In presenting the list of surgeons of the volunteer force which I have arranged in a tabular form, I take pleasure in acknowledging my indebtedness to Surgeon General Vanderpoel, for affording me every facility in its preparation, by allowing me free access to his official records, nor can I here forbear to speak of the very able and faithful manner in which the arduous and responsible duties of his office have been performed. Surgeon General Vanderpoel's early and persistent measures for the revaccination of the troops, (and his returns show the only statistics of revaccination in the present army,*) his energy in establishing hospitals at the general depots, his constant attention and watchfulness for the general health of the troops, his efforts to supply efficient surgeons to the various regiments, together with the minor duties, have been executed with indefatigable energy, and with an ability and patriotism that reflect credit not only upon our profession but on the State of New York.

* See Sanitary Commission Document E, page 26.

The Naval Medical Board, which convened in the early summer of 1861, and was dissolved by an order of the Secretary of the Navy, on the 27th of January, 1862, its duties being completed, accepted the following candidates:

NAMES.	Age.	Where graduated.
Adams, Newton H.		Albany Medical College.
Allingham, James J.		College of Physicians and Surgeons.
Brown, Wm. Mann		Buffalo University.
Brush, George R.		College of Physicians and Surgeons.
Clarke, Stephen H.		University of New York.
Covell, Charles E.*		University of New York.
Carter, Charles		College of Physicians and Surgeons.
Chalmers, William		College of Physicians and Surgeons.
Gunning, J. Henry		University of New York.
Hall, Watson C.		Geneva Medical College.
Lewis, F. B. A.		Harvard University.
Murphy, John D.		University of New York.
Plant, William S.		University of Michigan.
Tovey, Wm. F.		University of New York.
Smith, Heber		College of Physicians and Surgeons.
Storm, George D.		Buffalo University.
Zinzin, Lewis		McGill College, Montreal.

NOTE.—Sixteen of fifty-nine were from the State of New York.

* Charles E. Covell was born in Hartford, Washington Co., New York, in 1837. He was a student of Dr. J. Swinburne, in Albany, and was graduated in medicine at the University of New York in 1860. He became resident physician to the Children's Hospital, and subsequently became an interne at Bellevue Hospital, where, by his intelligence and fidelity, he gained warm friends. On the breaking out of the present war, he determined to enter the naval service, and was accordingly examined June 14th, 1861, before the Naval Board. He at once received orders, and sailed on the "CITY OF NEW YORK" on the 7th July for Pensacola. On the 22d July he was transferred by the United States flag ship MISSISSIPPI to the U. S. Frigate COLORADO, where he entered his duties as the assistant of Dr. Horer, the Surgeon of the fleet. He was in a few days seized with dysentery, of which he died on the 7th day of August, 1861. A few hours before he died, he penned a beautiful and affectionate letter to his mother, telling her of the fate he was in a few hours to meet, and of God's grace, which was sustaining him as the fatal hour approached. Thus was suddenly terminated a life full of professional zeal and promise. It was given to the cause of Liberty. Thus were crushed a mother's hopes in her only son. Such is the price of Liberty.

S. D. W.

The following gentlemen were candidates for the Medical Corps of the United States Army, from the State of New York, who were examined and approved by the Army Medical Board, convened in New York, May, 1861.

NAME.	Age.	Where graduated.	Year.
Bell, John..................	29	University of Pennsylvania	1854
Gray, Charles C.*.............	23	Hobart College, Geneva....................	1860
Gourley, John W. S.	29	College of Physicians and Surgeons........	1858
Goddard, Charles E...........	23	College of Physicians and Surgeons........	1859
Dunster, Edward Swift	27	New York Medical College	1859
Horton, Samuel M............	23	Jefferson Medical College	1861
Howard, Benjamin	25	College of Physicians and Surgeons..... ...	1859
Pooley, James H.	23	College of Physicians and Surgeons........	1860
Sprague, Havilah M..........	27	University Medical College, N. Y..........	1861
Spencer, William C.	23	Jefferson Medical College.................	1860
Steinberg, George M..........	23	do do do	
Temple, Joseph E.............	29	Jefferson Medical College.................	1853
White, Charles B.............	24	University Medical College, N. Y.	1859
Weir, Robert Fulton..........	23	College of Physicians and Surgeons........	1859
Winne, Charles K.............	23	Jefferson Medical College.................	1859
Beardsley, Grove S.†	23	Univerity of New York...................	1859

* "Would not leave his wounded." Prisoner at Bull Run and still unexchanged, June 5, 1862.

† Subsequently passed the Naval Board, and declined an appointment in the army.

The Sanitary Commission consists of the following gentlemen:

Rev. Henry Bellows, D. D., President, New York.

Prof. A. D. Bache, M. D., Vice President, Philadelphia.

Elisha Harris, Corresponding Secretary, New York City.

George W. Cullum, U. S. Army.

Alexander E. Shiras, U. S. Army.

Robert C. Wood, M. D., U. S. Army.

William H. Van Buren, M. D., New York.

Wolcott Gibbs, M. D., New York.

Cornelius R. Agnew, M. D., New York.

J. S. Newbury, M. D.

George T. Strong.

Subsequently added as commissioners under the authority of Secretary of War.

Frederick Law Olmstead.

Samuel G. Howe, M. D.

Horace Binney, Jr.

Rt. Rev. Thomas M. Clark, D. D., of Rhode Island.

Frederick Law Olmstead was appointed General Secretary at Washington.

J. Foster Jenkins, M. D., of Yonkers, New York, Secretary for the Army of the Potomac.

J. II. Douglas, M. D., of New York City, Secretary for the divisions under Generals Banks, Dix and Wool.

J. S. Newbury, M. D., Secretary for the western armies.

Franklin B. Hough, M. D., of New York, Secretary for the Army of the Potomac.

Surgeons who have been appointed to the charge of Brigades from the State of New York, and who were examined by the Medical Board in Washington.

NAMES.	Age.	Where assigned.
Azpell, Thomas S.		
†Bontecou, Reed B.*		General Wool's Division.
Brown, Rufus K.		
Burr, George		
Chamberlain, M. W.		
Church, W. H.		
Dalton, John C., Jr.		
Hamilton, Frank H.*		
Hewitt, Henry S.		
Hoff, Alexander H.*		General Hunter, Missouri.
Lidell, John A.		
McMillan, Charles		
Mosely, Nathaniel R.*		
McNulty, John*		General Dix's Division, Baltimore.
Oliver, George H.*		
Otterson, William C.		
Spencer, T. Rush*		
Strew, William W.		General Anderson, Kentucky.
Suckley, George		General Landers.
Thompson, J. Harry*		
Thurston, A. Henry		

* Previously examined by the Medical Board at Albany, and appointed to the charge of regiments.

† In charge of General Hospital at Fortress Monroe, Va.

Examined by the Medical Board at Albany, and passed as Surgeons New York State Volunteers.

NAME.	Age.	Where graduated.	Year.	What service since graduation.	Where appointed.	What changes.
Alexander, Joseph B.	38	University of Virginia	1852	Student in Charity Hospital; in practice 18 years		
Armstrong, Henry A.	42	New York Med. College	1854	Bartholomew Hospital 12 months; 3 years at Glasgow	Surg. 2d reg't artillery	
Atherly, Joseph	38	University of Glasgow	1851	In private practice	Surgeon 22d regiment	
Andrus, C. H.	38	College Phy'ns and Surgeons	1845			
Avery, A. Geer	40	Univ. of Louisville, Ky.	1845		Asst. Surg. Ma. arti'ry	
Barnes, Norman S.	35	Berkshire College	1852	3 months at Boston Hosp.; 3 months at New York Hosp.	Surg. 27th regiment.	
Backus, Azel	34	Univ. of Pennsylvania	1851	Had charge of Western House of Refuge 1 to 5 years		
Baker, Charles H.	43	Albany Medical College	1842	Hospital practice 1 year; in private practice 18 years		
Bannister, T. O.	28	University of New York	1856	House surgeon at Bellevue hospital 18 months; in private practice		
Barrows, N.	31	College Phy'ns and Surgeons	1856			
Bates, C. C.	33	New York University	1854			
Bausch, Charles	30	Tubingen	1851-7	Deputy coroner of New York		
Beach, John	30	College Phy'ns and Surgeons	1856	In private practice		
Beakes, George M.	30	Albany Medical College	1856	In private practice 25 years	Asst Surg. 1st reg't cav	
Benedict, Michael D.	47	Yale Medical Institute	1836		Surg. 75th regiment	
Berkey, Reese B.	34	Medical Dept. Penn. College	1849		Surg. Col. Doubleday's 1st reg. heavy artillery	
Berry, Abraham J.					Surg. 38th regiment.	
Bidlack, William W.	30	Michigan University	1852		Asst Surg. 62d reg't.	
Bissell, Elias L.	28	College Phy'ns and Surgeons	1861	In private practice	Asst Surg. 44th reg't.	
Bogardus, John C.	24	Castleton, Vermont	1860		Asst Surg. 102d reg't	
Bontecou, Reed B.	37	Geneva Medical College	1847	In private practice	Surg. 2d regiment	Promoted to Brig. Sur.
Bradley, Thomas	35	University Berlin	1847	Attended Hosp. in Paris and Germany		Chg Gen.Hos.F.Monroe
Brilliantosky, Samuel	37	College Phys. and Surgeons	1846	In general practice	Surg. 41st regiment.	
Bresee, J. R.	40	University New York	1856			
Brown, Harvey E.	25		1857		Asst Surg. 70th reg't.	
Browne, Rufus K.	27	Long Island College Hospital	1860	Resident Surg. L. I. Coll. Hosp.; Prof. Anatomy N. Y. Medical College.		Promoted to Brig. Sur.

SURGEONS.—Continued.

NAME.	Age.	Where graduated.	Year.	What service since graduation.	Where appointed.	What changes.
Brueninghausen, Charles ..	51	Berlin, Prussia............	1829	4 years in Marine Hospital; appointed by Gov. Fish........		Sur. 7th St'ben Rangers
Buck, Ephraim W........	29	College Phys. and Surgeons....	1857	Eastern Dispensary 1 year........	Surg. 51st regiment...	
Burr, William J........	43	Geneva Medical College......	1845	In private practice........	Asst. Surg. 59th reg't.	
Campbell, Augustus......	40	Albany Medical College	1845	Assistant physician at Bellevue 1 yr; private practice 15 years....		
Casey, James W	27	Buffalo University.........	1862	2 years interne St. Mary's Hospital	Surg. 77th regiment...	
Cass, Jonathan........	36	Albany Medical College......	1854	In general practice	Asst. Surg. 105th reg't	
Case, D. Meigs.........	25	New York University......	1857	Bellevue Hospital 5 months; House Surg. Eye Infirmary 6 months....		
Chamberlain, D. C. ..	46	Castleton, Vermont.......	1837	In private practice 24 years......	Asst Surg. 43d reg't ..	Promoted to Surgeon.
Chapin, Francis L. R....	36	Albany Medical College......	1851	Albany Hospital 3 months......	Surg. 105th regiment...	
Chapman, James	32	University New York	1852		Surg. 30th regiment...	
Chase, Enos G.	34	Jefferson Medical College...	1854		Surg. 8th reg. cavalry.	Resig'd, & A Surg 90th
Churchill, Alonzo......	50	Licensed by Ots. Co. Med Soc.	1832	In practice 23 years........	Surg. 104th regiment..	
Churchill, Seymour	43	Woodstock, Vermont......	1838	In general practice........	Surg. 14th regiment....	
Clark, A. M	26	College Phys. and Surgeons ...	1857	In general practi?e........	Surg. 23d regiment	
Cleveland, Warren......	41	Castleton, Vermont.......	1847	Been Health officer of Brooklyn; in private practice........		
Colgan, Joseph S........	42	Royal Coll. Surgeons, Dublin..	1849	Attended Dispensary in Brooklyn; also Hospital in Dublin........	Surg. 87th regiment...	Resigned.
Cooper, William S.......	42	Albany Medical College	1861	Practiced 19 years before graduation ...	Surg. 59th regiment ..	
Cotelle, Victor Amede ...	...				Asst Sur.Enfant Perdus	
Coventry, Walter B	27	Buffalo Medical College......	1858	Interne at Lying-in Hospital, Buffalo, 6 mo.; resident phy. Buffalo Hosp.	Surg. 26th regiment...	
Crandall, William B	27	Albany Medical College	1858	Hospital service 6 mo.: 2½ years in Hospital London and Paris........		
Cunynghame, Thurlow	28	McGill College, Montreal......	1858	House Surg. Hospital 2 years	Surg. 16th regiment...	
Cutter, J. L	33	College Phys. and Surgeons ...	1849	In private practice........	Surg. 101st regiment...	
Dalrymple, A. P.	36	College Phys. and Surgeons...	1851	1 yr at Bellevue; 2 yrs at St. Luke's Hospital	Surg. Eng. & Art. reg.	Resigned.
Dalton, Edward B.......	26	College Phys. and Surgeons ...	1858		Surg. 36th regiment....	
Danaghe, William R.	31	New York University......	1852	18 mo. Asst. Surg. at Bellevue; 6 yrs. Surg. Demilt Dispensary, &c....		
David, William G	30	Harvard Medical School	1854	9 mos in New York City Hospital....	Surg. 98th regiment...	

Name	Age	College	Year	Previous service / experience	Assignment	Remarks
D'Avignon, Francis J.	54	Montreal, L. C.	1833		Surg. 96th regiment	Promoted to Surgeon.
Dexter, James E.	37	New York University	1860		Asst. Surg. 40th reg't.	
Dickerson, D Estaing	27	Albany Medical College			Asst. Surg. 33d reg't.	
Dilts, George S.	30	New York University	1856		Surg. Jackson artillery	
Dimon, Theodore	44	University of Pennsylvania	1838	Hospital practice in France, Dublin, &c Phy., in Auburn State Prison 2 yrs; in private practice	Surg. 19th regiment	
Douglas, Frederick	25	New York University	1859	Attended Bellevue and N. Y. Hosp. 6 mos; attend. Phy. Rochester Hosp. 1 year	Asst. Surg. 3d reg. env.	Promoted to Surgeon.
Dunmainville, Lucian	22	Long Island Hospital	1860	Asst. of Dr. Rochester at Buffalo Hosp.	Asst Surg. 31st reg't	
Dunster, Edward L.	27	New York Medical College	1859	St. Luke's Hosp. 6 mos; acting Surg. Demilt Disp. 2 yrs	Asst. Surg. U. S. A	
Ellis, Thomas T.	39	Royal College Surgeons. Lond. Royal College Phys., England	1845		Surg. 7th regiment	
Eisenlord, A. M. F.	36	New York University	1844			
Everts, Franklin	31	College Phys. and Surgeons	1854	At Bellevue Hosp. 1855; Hosp. Paris and London 1856	Surg. 1st reg't artillery	Resigned, and app'ted Surg. 8th cavalry.
Ferguson, Nelson D.	41	Geneva Medical College	1844	In private practice nearly 18 years	Surg. 97th regiment	Consolidated to 95th and must. out service.
Fisher, George J.	36	University New York	1849	Surgeon Sing Sing prison, Priv. Prac.	Sur.3d reg. Eagle Br'de	Resigned.
Fiske, S. N.	24	New York University	1860	7 yrs Surg. British army; several yrs. in Hosp s of Europe	Surg. 25th regiment	Deceased.
*Foy, M. Eyre	37	L. R. C. of Edinburgh	1846		Surg. 38th regiment	
Froehlick, Mority A	31	Tubingen New York Medical College	1854 1856	In private practice	Surg. 8th regiment	
Frothingham, William	31	College Phys. and Surgeons	1854	Bellevue Hosp. 2 yrs	Surg. 44th regiment	
Glennan, Patrick	35	Geneva Medical College	1850			
Goodale, Charles	44	Albany Medical College	1841		Surg. 94th regiment	Resigned.
Goodrich, Charles S.		College Phys. and Surgeons	1829	Prof. Anatomy N. Y. Med. Coll.; 10 yrs Hosp. practice	Surg. 102d regiment	Resigned.
Gouley, J. W.	29	College Phys. and Surgeons	1853	In general practice		
Gibbs, T. C.	34	Castleton, Vermont	1853	In general practice; general practice	Surg. 5th regiment	Promoted to Brig. Sur
Gilbert, Rufus H.	29	College Phys. and Surgeons	1852	1 yr in army South Florida; 1856 Surg. on steamer; general practice	Sur. 78th reg.	
Gray, E. P.	37	University of Buffalo	1849	Private practice	Surg. 7th reg't cavalry	
Haak, John William	45	Saxony	1839	Hospital practice 3 years	Surg. 49th regiment	
Hinekley, Charles E.	25	University of Pennsylvania	1859	Private practice 3 years		
Hall, James A.	45	Bowdoin, Maine	1838	Asst. Dem. when grad.; in gen'l prac.	Surg. 54th regiment	
Haerle, F.	26	Tubingen, Wurtzemberg	1859			
Hagen, Charles W.	32	St. Louis, Medical College	1861		Surg. 54th regiment	
Halsey, C. E.	27	College Phys. and Surgeons	1858	L. Island Hosp. 1 yr; in private prac	Asst. Surg. 40th reg't.	

SURGEONS.—Continued.

NAME.	Age.	Where graduated.	Year.	What service since graduation.	Where appointed.	What changes.
Hamilton, Frank H.		......	......	Hosp. Vienna,Wurzburg, Tubingen, &c	Surg. 31st regiment	Promoted to Brig. Sur.
Hausen, Julius	33	Vienna, Austria	......	House Surg. Bellevue Hosp. 1 yr; ship	Surg. 20th regiment	
Hunt, John W.	27	New York University	1859	Surg. 1 voyage	Surg. 10th regiment	Promoted to Brig. Sur.
Hoysradt, J. H.	30	Albany Medical College	1853	In general practice	Surg. 10th regiment	
Heller, T. Lewis	36	Heidelberg, Freyburg	......	......	Surg. 103d regiment	
Heiland, Charles	28	Gottingen, Kingsven	......	Surgeon of steamship "Hudson Wiscr" and "New York," &c	Asst. Surg. 20th reg't	
Helmer, Albert M.	29	University of Michigan	1858	Visitor at Bellevue and N. York Hosp. for 8 mos	Surg. 28th regiment	
Hermann, August	35	College Phys. and Surgeons	1858	......	Asst. Surg. 29th reg't	
Hessell, Ferdinand	28	Marburg, Hesse	1856	......	Surg. 58th regiment	Resigned.
Hewett, James D.	27	New York University	1858	In private practice 3 years	Asst. Surg. 66th reg't	Resigned.
Hewitt, Henry	64	Medical Dept. Yale College	1826	......	Surg. 92d regiment	
Hicks, J. Lawrence	26	College Phys. and Surgeons	1860	House Surg. at Bellevue 1 year	Surg. 1st regiment	Resigned.
Hoff, Alexander H.	39	Jefferson Medical College	1845	Hosp. service some years	Surg. 3d regiment	Promoted to Brig. Sur.
Hoffmann, Edward S.	31	College Phys. and Surgeons	1855	......	Surg. 90th regiment	
Hubbell, Charles L.	34	Berkshire Medical College	1848	In private practice 7 years; attending Surgeon Troy Hospital		Reg. disban'd and app. Sur. 12th reg't inf.
Hovet, Henry	35	Gottingen, Germany	1852	......	Surg. 2d reg. B.H.cav	
Howard, Benjamin	26	College Phys. and Surgeons	1857	House Surg. Jews' Hosp. New York; Ship Surg. Atlantic line	Surg. 46th regiment	
Howe, John	22	New York Medical College	1859	1 year House Phy, at Bellevue, &c	Asst. Surg. 19th reg't	Resigned & app.U.S.A
Humphreys, George H.	26	Jefferson Medical College	1856	Student 2 yrs in Hosp. Vienna, Berlin, Paris	Asst. Surg. 1st reg't	Promoted to Surgeon.
Hutchinson, William F.	24	New York University	1858	Army service in Charleston harbor; in Charity Hosp., New Orleans	Surg. 9th regiment	
Ideler, Herman	42	School of Berlin, Prussia	1844	Hospital practice in Berlin	Asst. Surg. 22d reg't	
Irwin, Charles K.	36	Albany Medical College	1855	In practice since grad	Surg. 45th regiment	
Jamison, John S.	39	University of Michigan	1852	In private practice	Surg. 72d regiment	
Jenkins, J. Foster	35	University of Pennsylvania	1848	Visitor at Mass. General Hosp. 6 mos	Surg. 86th regiment	
Joachim, Conrad	43	Wurtzberg			Sanitary Commission	
Jungbanns, L. H.	28	Munich, Wurtzberg, &c	1856	Asst. at Lunatic Asylum, Flatbush, 1 year, &c	Asst. Surg.3d Bat.Art.	
Jules Debreuil	36	......	......	Externe and interne of Paris Hosp'ls		

Name	Age	Medical education	Year	Service record	Army rank	Remarks
Kelsey, Dana E.	26	Albany Medical College	1857	Attending Hosp. New York and London	Asst. Surg. 64th reg't.	
Kinmier, James	55	University Glasgow	1822	Passed by Royal Coll. Phy., London; competent for British service		
Kittinger, Martin S.	34	College Phys. and Surgeons	1853		Surg. 100th regiment	
Kneeland, Jonathan	49	Licensed by Onon. Co. Med. So.		In private practice	Volunteer Surg.	
Knight, William W.	28	Berkshire Medical College	1855	In general practice	Asst. Surg. 19th reg't now 3d artillery	
Koehler, Augustus	37	Wurzburg, Prague		Surg, Austrian army 4½ years; in practice 8 years		
Leach, George H.	39	Castleton, Vermont	1846	In practice—chiefly surgical	Surg. 57th regiment	Resigned.
Legler, Henry T.	40	Leipsic, Germany	1844		Asst. Surg. 8th reg't.	
Lewis, John D.	32	Albany Medical College	1854	4 mo. Albany Hosp.; 3 mo. Buffalo Hos.	Asst. Surg. 85th reg't	
Little, David	27	College Phys. and Surgeons	1858	Asst. and House Phy. Bellevue, 18 mo.	Surg. 13th regiment.	
Little, William B.	39	Bowdoin College, Maine	1860		Surg. 32d regiment.	
Lyman, William C.	25	Berkshire Medical College	1859		Asst. Surg. U. S. navy	
Lynch, Edmund	24	Long Island Hospital	1860	Asst. at Long Island Hosp. 1 year	Asst. Surg. 6th reg't.	
Lovejoy, George W.	24	College Phys. and Surgeons	1859	Surgeon Eastern Dispensary 18 mos.	Asst. Surg. 4th reg't.	
Major, Adolph	40	University Heidelberg	1843	Military service in 1848, '49, in Germany; in practice 12 years		
Mansfield, Wm. Q.	42	Buffalo University	1847	In general practice	Surg. 4th cavalry.	
Marguerat, E.	32	New York University	1859		Asst. Surg. 92d reg't.	
Martin, B. Ellis	23	College Phys. and Surgeons	1561			Resigned.
Martindale, Frank E.	30	Albany Medical College	1852	Dep. Health officer at Quarantine 2 yrs.	Asst. Surg. 5th reg't.	
Marvin, L. J.	55	Fairfield Medical College	1827		Surg. 97th regiment.	
Mattimore, Frank J.	27	Albany Medical College	1860	Asst. 1 yr at Albany Co. Hosp. Asylum		
May, Henry C.	30	Michigan University	1856	In practice 5½ years	Surg. 5th regiment.	
McDermott, Wm. J.	30	New York University	1853	Surg. 3 mos. 6th regiment N. Y. S. M.; attended Hosp. New York and Paris	Surg. 66th regiment.	
McDonnell, Edward	39	College Surgeons, London / College Physicians, London	1842 / 1845		Asst. Surg. 1st Artil.	
McKay, Lawrence	31	Buffalo Univ. Med. College	1850	In practice 6 years	Surg. 6th reg't cavalry	
McLean, Le Roy	30	Albany Medical College	1855	1 year Asst. House Physician Albany Hosp.; 4 years Marshall Infirmary		
McNair, James	45	Lond. Coll. Phys. and Surgeons	1839	In priv. practice 11 yrs; in Ireland 10 yr	Asst. Surg. 2d reg't.	Promoted to Surgeon.
McNulty, John	36	New York University	1853		Surg. 15th regiment.	Promoted to Brig. Sun
McNutt, Hiram	41	New Hamp. Medical Institute	1845	In practice 16 years	Surg. 37th regiment.	
Mercer, John T.	32		1855			
Merrill, Andrew	35	Geneva Medical College	1850	In private practice	Asst. Surg. 61st reg't.	Resigned.
Metcalfe, George W.	35	New York University	1857		Asst. Surg. 76th reg't.	
Morris, Robert	45	Albany Medical College	1846	In private practice 15 years	Surg. 91st regiment.	
Moseley, Nathaniel R.	35	University of Pennsylvania	1847	2 years in charge Med. Dept. Blockley Hospital, &c.	Surg. 36th regiment.	Promoted to Brig. Sur.

SURGEONS.—Continued.

NAME.	Age.	Where graduated.	Year.	What service since graduation.	Where appointed.	What changes.
Moses, Israel..........	37	College Phys. and Surgeons....	1845	Surgeon in U. S. A. 1846 to 1855; 1 year in Paris Hosp........	....	Lieut. Col. 72d reg't.
Mucke, Franz.........	28	Greifswald, Prussia..........	1856	4 years Assistant Surgeon in Hospital.	Asst. Surg. 58th reg't.	Promoted to Surgeon.
Mudie, Archibald F	25	College Phys. and Surgeons...	1860	....	Asst. Surg. 15th reg't.	Resigned.
Mulford, Sylvanus S....	31	College Phys. and Surgeons...	1855	....	Asst. Surg. 35th reg't.	Promoted to Surgeon.
Murdock, James B........	31	College Phys. and Surgeons...	1854	House Surg. at Bellevue 2½ yrs; Surg. on steamer......	Surg. 21th regiment ..	
Nelson, Judson C.	36	Geneva Medical College.....	1845	In practice 18 years......	Surg. 76th regiment...	
Nawhaus, Charles Ed.....	41	Berlin................	1846	Asst. Surg. Prussian army; visited 4 years Hosp. Berlin, Germany......	Surg. 29th regiment...	
Nordquist, Charles J.	39	New York University	1854	4 years in Hosp. of Europe; 7 years in practice........	Surg. 83d regiment ...	
Oliver, George H........	29	Harvard University........	1857	2 years St. Vincent Hospital........	Asst. Surg. 37th reg't.	Promoted to surgeon.
O'Meagher, William	29	New York University	1844	In private practice ....:.....	Surg. 3d reg't cavalry.	
Palmer, John M........	40	Licensed by County Med. Soc.	1858	....		
Palmer, William H.	31	New York University				
Parsons, William W. D....	29	New York Medical College	1853	In private practice......	Surg. 6th regiment.....	
Pease, Philo C..........	26	Long Island Hospital......	1860	Resident Surg. at Long Island 1 year.	Surg. 12th regiment...	Resigned—app. Surg. 10th reg't cavalry.
Pease, Roger W..........	34	Geneva Medical College	1847	Baltimore Infirmary 7 years 3 months.		
Perkins, E. T.	34	College Phys. and Surgeons.....	1856	Surg. Atlantic and Isthmus steamers.	....	
Perry, John L..........	46	Vermont Medical College......		In general practice 25 years.....	Surg. 77th regiment ..	Resigned.
Petherbridge, John D.....	34	University of Pennsylvania.....	1847	....	Surg. 65th regiment ..	
Phillips, Henry John......	31	College Surgeons. London / St. Andrews, Scotland	1849 / 1857	On Med. staff thro' Crimean war, '54	Surg. 53d regiment......	{ Reg. dis'd., app. surgeon 102, and res'd.
Plumb, S. Hiram........	42	Licensed by Way. Co. Med. So.	1844	In private practice......	Asst. Surg. 24th reg't.	
Potter, Hazard A........	48	Bowdoin College, Maine......	1835	Visited Mass. and Philadelphia Hosp.	Surg. 50th regiment ..	
Powers, Cyrus..........	44	Geneva Medical College	1845	Attending Hosp. Buffalo and New York.	....	
Reisberg, Henry W.......	27	New York Medical College	1860	In private practice......	Asst. Surg. 75th reg't.	
				Visited Hosp. in England and America 7 years	Asst. Surg. 68th reg't.	
Rouss, P. Jos...........	36	....	1850			
Reynolds, Francis.........	30	Dublin	1849	Late Asst. Surgeon British army, in Crimea........	Surg. 88th regiment...	
Reynolds, Lawrence.......				....	Asst. Surg. 24th reg't.	Promoted Sur. 63 reg.

Name	Age	Medical School	Year	History	Service	Remarks
Rice, Nathan P.	31	Harvard Univ. Med. Dept.	1852	3 years visited Hosp. of Europe; House Surg. Mass. General Hospital	Surg. 18th regiment.	
Rice, Pitkin B.		Harvard Univ. Med. Dept.	1853	Been Surg. 28th regiment 3 mo's Vol.	Surg. 81st regiment.	
Rice, William H.	40	College Phys. and Surgeons	1850	Hosp. practice 18 mo.; priv. prac. 11 yrs		Deceased.
*Ruggles, Eli Samuel	32	Edinburgh	1861			
Sabin, S. A.	31	University of Michigan	1857	1 year Hospital in Detroit		
Sass, Louis	35	Copenhagen	1847			
Schenck, Otto	35	Giessen	1848	Visited for 3 years Hospital of Glasgow	Asst. Surg. 46th reg't.	
Sheldon, Andrew F.	31	New York University	1852		As. Surg. 2d reg't cav.	Dis'd, asst. surg. 78th.
Sherman, Socrates N.	59	Middlebury College, Vermont	1824	In active practice 37 years; Member of Congress	Surg. 34th regiment.	
Shipman Azriah B.	55	{ Castleton, Vermont / Jefferson Medical College }	1836 / 1842	Prof. Surgery Indiana Med. College 8 years	Asst. Surg. 17th reg't.	{ Pro. to Surg. 12th re; promo. to Brig. Surg.
Shultze, Louis	38	Burlington	1846	Asst. Prof. Surg. N. Y. Med. Coll. '61	Surg. 68th regiment.	
Shultze, Augustus		St. Louis	1860	Asst. Surg. Sisters' Charity Hospital, St. Louis		
Simon, John	32	Gottingen, Hanover	1857			
Sloat, Spencer S.	33	College Phys. and Surgeons	1849		Surg. 95th regiment.	Promoted surg. 94th.
Smith, Andrew H.	24	College Phys. and Surgeons	1858		Asst. Surg. 43d reg't.	
Smith, Joseph T.	36	Jefferson Medical College	1854	In private practice		
Smith, J. Paschal	28	New York Unversity	1858	Surg. 69th militia regiment; served on Blackwell's Island Hospital staff	Surg. 69th regiment.	
Smith, Strowbridge	33	New York Medical College	1851	House Surg. at Ward's Island; State Emigrant office 1853 to 1855	Surg. 93d regiment.	
Smith, William A.	41	Genera Medical College	1847	In practice 15 years	Asst. Surg. 89th reg't.	
Smith, William M.	36	Castleton, Vermont	1840	In practice except two winters	Surg. 85th regiment.	
Snow, Asa B.	52	Fairfield Medical Col. N. Y.	1832	In private practice	Surg. 61st regiment.	Discharged; app. Surg. Serrell's Engr. reg't.
Spencer, Henry T.	34	Albany Medical College	1852	1 year Hosp. in Albany; 2 years Alms House		
Spencer, T. Rush	43	University of Pennsylvania	1840	Resident Phy. at Blockley Hosp.; 1 year Prof. Genera Medical College		Promoted to Brig. Surg.
Squire, Truman H.	38	College Phys. and Surgeons	1849		Surg. 33d regiment.	
Stachle, Francis R.	33	Bern, Switzerland	1858		Surg. 89th regiment.	
Stearns, Charles W.	42	University Pennsylvania	1840	Served Mass. Gen'l Hosp.; commissioned Medical staff 1841	Asst. Surg. 7th reg't.	
Stebbins, Roderick	44	Berkshire Medical College	1842	In private practice	Surg. 3d regiment	
Steele, Aaron J.	24	Jefferson Medical College	1859	Res. Physician Buffalo Hospital; Dem. of Anatomy Buffalo Med. Coll. 14 m's		
Stein, L.	50	{ Tubingen / Wurzburg }	1831 / 1833		Asst. Surg. 26th reg't.	
Steinach, Adelrick	33	St. Gall	1854	Att. Hosp., Paris; in practice in N. J.		

SURGEONS.—Continued.

NAME.	Age.	Where graduated.	Year.	What service since graduation.	Where appointed.	What changes.
Stephenson, Mark	47	College Phys. and Surgeons	1826	2 yrs New York Hosp.; 8 yrs New York Dispensary; 9 yrs Ophthalmic Hosp.	Asst. Surg. U. S. army	
Sternberg, George M.	23	College Phys. and Surgeons	1860	In active practice	Surg. 52d regiment	
Stiebeling, George C.	30	Giessen and Marburg	1854	In practice 6½ years		
Stow, T. Dwight	32	Cleveland College, Ohio	1854	In active practice		
Strong, Thomas D.	39	Buffalo University	1851			
Stuart, James C.	48	Berkshire Medical College	1835	Hospital practice Boston and N. York	Surg. 17th regiment	
Terry, W. F.	27	New York University		Hosp. practice 15 mos; general practice 3 years	Surg. 40th regiment	Res; app. Sur. 43d reg; pro. Brig Surg., dis'd
Thompson, J. Harny						
Tice, Lewis	34	New York University	1851		Asst. Surg. 17th reg't.	
Tingley, William H.	37	Jefferson Medical College	1846		Asst. Surg. U. S. army	
Torrey, Charles W.	25	University of Wurzburg	1858		Asst. Surg. 51st reg't.	Dismissed the service.
Totten, Gilbert T.	21	College Phys. and Surgeons	1861		Asst. Surg. 32d reg't.	
Townsend, Morris W.	34	Jefferson Medical College			Surg. 47th regiment.	
Trenor, Eustace	27	College Phys. and Surgeons	1853			
Van Benst, Bernard	30	Leipsic	1856	Served 2 mos as Sub. In N. York Hosp.		
Vanderkieft, B. A.	54				Asst. Surg. 53d reg't.	Disb'd, surg. 102d reg.
Vaudrey, E.		Medical School Paris	1836	Surg. to poor in Paris from 1836 to '52	Surg. Enfant Perdus.	
Van Etten, Solomon	31	Albany Medical College	1855	In active practice	Surg. 56th regiment.	
Van Ingen, James L.	43	College Phys. and Surgeons	1843		Surg. 18th regiment.	Resig'd; app. Surg.5th reg't; subseq'ly dis'd.
Van Slyke, David B.	33	Buffalo Medical College	1852	In practice 9 years	Asst. Surg. 101st reg't	
Van Slyke, Dewitt C.	40	Geneva Medical College	1850	In active practice	Surg. 35th regiment.	
Walker, Edward S.	37	New York University	1849	In active practice	Surg. 34th reg't.	
Wainwright, D. Wadsworth	27	College Phys. and Surgeons	1854	In active practice	Surg. 4th regiment.=	
Weed, Hiland A.	27	New York University	1857	In active practice	Asst. Surg. 17th reg't.	Pro. to Surg. 25th reg.
Welcker, Rudolph	40	University Germany	1845	1st Asst. Surg. of University Hospital.	Surg. 8th regiment.	Promoted to Brig. Surg.
Welles, Samuel R.	36	Buffalo Medical College	1848	In private practice	Surg. 61st regiment.	
Whedon, George D.	29	Albany Medical College			Asst.Surg.10th regt cav	
White, John P. P.	23	College Phys. and Surgeons	1861	Asst. Surgeon Bellevue Hosp	Asst. Surg. 9th reg't.	Promoted to surg. 10th
White, Whitman V.	27	{ Berkshire Medical College / New York Medical College	1857 / 1858	{ 1 yr in Blackwell's Is. Hosp.; In prac.	Surg. 47th regiment.	Resigned.

Wieber, George........	36	Stud in Halle, Wurtzb'g Giessen	1857	Served in Prussian army........	First Battalion artillery
Wilcox, Charles H......	....	....	....	....	Surg. 21st regiment...
Wilson, James........	25	Dublin	1855	In practice 6 years.......	Surg. 99th regiment...
Winne, Charles K......	23	Jefferson Medical College....	1854	Surg. in Buffalo Hosp. Sisters of Charity	Asst. Surg. U.S. army
Wolf, Frederick	33	Prague, Austria	1853	....	Surg. 39th regiment...
Wood, Charles S.......	35	Jefferson Medical College....	1851	....	Asst. Surg. 66th reg't.
Wood, Lucian P.......	31	Berkshire Medical College....	1854	....	Asst. Sur. 5th regt.cav
Wright, F. Markoe......	32	Coll. Physicians and Surgeons..	1854	3 years asst. State Lunatic Asylum, and private practice.......	Asst. Surg. Col. Dodge mounted rifles.
Wunderlick, Gerald.....	56	....	1828	Practiced in Hosp. at Wurtzburg 26 yrs	....
Wylie, Farand........	42	Geneva Medical College	1847	....	Asst. Surg. 86th reg't.
Young, Oscar H.......	25	Albany Medical College......	1858	House Surg. and Phy. at Albany Hosp.	....

† Wounded at Hanover C. H. battle.

4

Examined by the Medical Board at Albany, and passed as ASSISTANT SURGEONS New York State Volunteers, 1861.

NAME.	Age.	Where graduated.	Year.	What service since graduation.	Where appointed.	What changes.
Allen, Isaac B......	47					
Allingham, James J...	25	College Phys. and Surgeons	1859			
Avery, George W......	34	Albany Medical College	1850	Asst. Surg., Rochester Hospital; practised in Rochester	Ass't. Surg. 13th reg't, "Rochester reg't."	
Ayme, H......	43	University of Pennsylvania		12 years in practice		
Bacon, James G....	21	Albany Medical Coll.	1860			
Baleh, Galusha B....	22	College Phys. and Surgeons	1860	In N. Y. and Bellevue Hospital	Ass't Surg. 98th reg't.	
Barron, John C. ...	23	College Phys. and Surgeons	1861			
Bates, Newton L......	24	Jefferson Medical College	1861	7 months in Buffalo Hospital; resident physician Erie Co. Hospital		
Bayles, George......	25	College Phys. and Surgeons	1859		Ass't Surg. U. S. N. / Ass't Surg. Col. Doubleday's artillery.	
Beardsley, Grove S....	23	New York University	1859	House Asst. Surgeon at Bellevue for a few weeks	Ass't Surg. U. S. N.	
Benedict, A. C......	24	Yale Medical College	1860	In private practice		
Bennett, George C. ...	34	Cleveland Medical College	1851		Ass't Sur. 9th reg. cav.	Resigned.
Blauvelt, J. F......	38	College Phys. and Surgeons	1851		Ass't Surg. 5th art'y.	
Bogert, E. S.	25	New York University	1860	In Bellevue Hospital		
Bogert, R. D.	22	College Phys. and Surgeons	1861			
Bradford, Theron ...	36	Indiana Medical College	1858			
Brown, D. M....	23	Cleveland Medical College	1860			
Brown, Spencer H...	30	New York University	1857		Asst Sur. 2d reg. artil.	Resigned, app. ass't surgeon U. S. Navy.
Brown, Edward A. ...	42	Mennsingen, Bavaria.	1849		Ass't Surg. 31st reg't.	
Braun, Martin....	26	College Phys. and Surgeons	1861			
Bryce, William M...	24	Albany Medical College.	1859	In Albany Hospital 10 months.		
Burbeck, Charles H.	31	College Phys. and Surgeons	1852	In private prac. 7 yrs.; 3 yrs. Hosp. prac.		
Burdett, Abram S....	30	Worcester, Mass.	1853	In private practice.		
Burdick, J. T....	36		1853			
Caceaun, Joseph ...	37	Woodstock	1848	One year in Pennsylvania Hospital.		
Cato, H. J. ...	34	Castleton, Vermont.	1855		Ass't Surg. 56th reg't.	
Carroll, O. A. ...	28	Harvard Univ. Med. Col.	1861		Ass't Sur. 6th reg. cav.	
Clark, Augustus P. ...	25	Albany Medical College.	1858	In private practice.	Ass't Surg. 60th reg't.	
Chambers, Wm. B. ...	47	Woodstock	1837			
Chase, Nathan B......						

Name	Age	College	Year	Experience	Appointment	Remarks
Cochrane, A. H.	27	Albany Medical College	1857	Spent 2 years in Hospital of Paris		Promoted to Surgeon.
Cooper, John	26	University of Pennsylvania	1857	In Dutchess Co. Hospital 7 or 8 years	Ass't Sur. 5th reg. cav.	
Cooper, John R.	34	College Phys. and Surgeons	....		Ass't Surg. 97th reg't.	
Cornish, Aaron	30	Castleton, Vermont	1854	In N. Y. Hospital on surgical side		
Cutler, G. K.	21	College Phys. and Surgeons	1860		Ass't Surg. 57th reg't.	
Dean, Henry C.	23	Harvard Univ. Med. College	1861		Ass't Surg. 84th reg't.	
Dewey, David B.	27	New York University	1858		Ass't Surgeon Baker's California regiment.	
Dwinelle, Justin	39	Jefferson Medical College	1846	In private practice 16 years		
Dodge, John L.	36	New York University	1845	6 months drug store; 6 months ship surg.; 6 months camp Cal.	Ass't Surg. 51st. reg't.	Promoted to Surgeon.
Doolittle, Frank W.	24	College Phys. and Surgeons	1860	Resident Surg. at Jews Hospital; Resident Physician at Child's Hospital	Ass't Surg. 10th reg't.	
Douglas, George C.	27	Albany Medical College	1857	In private practice	Ass't Surg. U. S. A.	
Downing, J. C. C.	23	Pennsylvania Medical College	1859	2 yrs. Surg. emigrant packet ship		
Duane, Henry	..	Albany Medical College	1861			
Edmonston, Alex'r A.	31	Albany Medical College	1853	Visited New York Hosp. in 1857 and 1858.	Ass't Surg. 18th reg't.	
Elliott, Samuel R.	25	New York Medical College	1856	Externe Hospitals, Paris, 1860, &c.		
Elting, V. V.	33	Berkshire Medical College	1851	Practiced 5 yrs. in Greene Co., balance in Hospital		
Feldbausch, Phillip	29	Wurzburg	....	Visited Hospital Wurzburg, Munich, Berlin, Paris, &c.		
Forrester, James Jr.	26	College Phys. and Surgeons	1858	Phys Northern Dispensary, New York	Ass't Surg. 45th reg't.	
Fossard, George H.	23	Albany Medical College	1859	In private practice 3 years	Ass't Surg. 42d reg't.	
Fuller, Winfield S.	23	College Phys. and Surgeons	1861		Ass't Sur. 8th reg. cav.	
Franklin, Morris J.	30	New York University	1858	Visitor at Bellv. and N. Y. Hosp. 3 years.	Ass't Sur. 4th reg. cav.	
French, Seth	37	Castleton, Vermont	1847	In priv. prac.; in Hosp. Sacramento City.	Ass't Surg. 35th reg't.	
Gale, James L.	33	Castleton, Vermont	1853	In priv. practice 9 years	Surgeon 50th regiment	
Gessner, Brower	26	University of Pennsylvania	1856	Served one year in South Carolina	Ass't Surg. 38th reg't.	
Gilligan, Michael G.	39	New York Medical College	1851	In priv. prac.; atten'd Hos. Dublin & N. Y.	Ass't Surg. 63d reg't.	
Goodrich, B. F.	21	Cleveland Medical College	1861			Dismissed.
Gradenlorff, Herman	37	New York Medical College	1854	In private practice		
Grimes, F. S.	27	Berkshire Medical College	1858			
Griswold, Stephen	33	New York University	1850	Assis't. Ward's Island Hosp. one year	Ass't Surg. 38th reg't.	Taken pris. at Bull Run Died at Charleston, typ'd fov. Nov. 1861.
Haldon, James	29	New York Medical College	1861			
Hall, William H.	36	Albany Medical College	1856		Ass't Surg. 36th reg't.	
Haynes, Jonathan K.	25	Albany Medical College	1857			
Hewett, Charles N.	41	Albany Medical College	1847	In private practice	Ass't Surg. 50th reg't.	
Holden, Austin W.	61	Dartmouth Medical College	1825	In private practice	Capt. 22d regiment	

ASSISTANT SURGEONS.—Continued.

NAME.	Age.	Where graduated.	Year.	What service since graduation.	Where appointed.	What changes.
Humphries, Patrick H.	26	New York University	1861	Attending physician in Williamsburg Disp.	Ass't Surg. 48th reg't.	
Ingerson, H. H.	27		1860	In general practice		
Isham, Nelson	50	Yale College Med. Dept.	1828	Phys. Herkimer Co. poor house 6 years		
Jackh, Gottlieb	30	Basle	1854	Attended Hospital at Basle, &c.	Ass't Surg. 7th reg't.	Resigned.
Johnson, William E.	28	Albany Medical College	1859	In private practice		
Kayner, D. S.	36	Castleton, Vermont	1849	In private practice		
Kidder, Walter	38	Berkshire Medical College	1845	In private practice		
Kilmer, Washington	23	Albany Medical College	1860	1 yr. as Ass't at Alms House Hospital		
Kinne, William B.	46	Berkshire Medical College	1835	In general practice	Ass't Surg. 90th reg't.	Resigned.
Kipp, Charles J.	23	College Phys. and Surgeons	1861	Ass't Surg. 5th Reg. N. Y. S. M. 3 mos.	Assistant Surg. 3d battalion.	Resigned.
Kneuten, John	28	Marburg, Wurzburg and Vienna.	1860			
Lakeman, William H.	25	St. Thomas Med. and Surgical College, London	1857	Hospital steward		
Landon, Douglass S.	35				13th regiment	
Lawyer, Thomas	30	Albany Medical College	1851	In private practice	Ass't Surg. 104th reg't.	
Leighton, Nathan'l W.	28	New York Medical College	1858	One year Kings Co. Hospital	Ass't Surg. 43d reg't.	
Lewis, John B.	29	New York University	1853	In private practice	Ass't Surg. 72d reg't.	
Little, George W.	25	Albany Medical College	1858	6 months Albany City Hospital		
Long, Alfred J.	36	New York University	1853	In private practice 8 years		
Mackay, David	26	Glasgow, Scotland	1856			
Madill, William A.	27	Albany Medical College	1858	Ass. Surg. Pennsylv'a Hospital 6 months, Physician and Surgeon Alms House	Ass't Surg. 23d reg't.	
Mann, William B.	23	Buffalo University	1861	N. Y. Disp. 4 years; N. Y. Eye and Ear Infirmary 2 years		
Marshall, Benjamin	30	College Phys. and Surgeons	1852	Assistant Physician Randall's Island Hosp.		
Marshall, E. G.	23	College Phys. and Surgeons	1861	Hospital		
Macfarlane, Carrington	25	College Phys. and Surgeons	1861	N. Y. City and Bellevue Hospital	Ass't Surg. 81st reg't.	
McAllister, Thomas	33	College Phys. and Surgeons	1853	In private practice 13 years		
McLellan, F. M.	41	Harvard University	1822	In private practice	Ass't Surg. Mar. Artil.	Promoted to Surgeon.
McGowan, John J.	40	Harvard University	1847	In private practice		
McKim, Robert V.	21	New York Medical College	1861		Ass't Surg. 57th reg't.	Promoted to Surgeon.
McKee, J. G.						
Mead, Martin L.	27	Albany Medical College	1859	Albany Hospital		
Michel, Louis	34	New York Medical College	1857	In private practice		

Name	Age	Medical College	Year	Experience	Assignment	Remarks
Miller, Adam	42	Geneva Medical College	1844			
Moore, J. W.	23	Castleton, Vermont	1858			
Mooers, John H.	33	New York University	1857		Ass't Surg. 16th reg't.	
Morse, Burnett W.	23	College Phys. and Surgeons	1860	Visitor at Bellevue and City Hospital	Ass't Surg. 27th reg't.	
Mudge, Charles	29	College Phys. and Surgeons	1854		Ass't Surg. Engineer & Artisans regiment.	
Mullen, Isaac V.	34		1851			
Murphy, Daniel H.	27	Buffalo Medical College	1856	2 years Hospital of Buffalo	Ass't Surg. 25th reg't.	Resigned.
Murray, William D.	27	National Coll. Wash., D. C.		In private practice 2 years 8 months	Ass't Surg. 100th reg't	
Myers, John T.	23	Albany Medical College	1859	8 months Blackwell's Island Hospital	Ass't Surg. 91st reg't	
Neely, Nelson	27	Albany Medical College	1854			
Norris, Thomas P.	33	New York University	1853	Attended Bellevue Hospital		
O'Leary, Cornelius B.	22	Albany Medical College	1860	Was Surgeon 25th Regiment N. Y. S. M. 3 months service		
O'Neil, James C.	29	New York University	1857		Ass't Surg. 25th reg't.	Resigned.
Osborne, Charles H.	24	College Phys. and Surgeons	1858		Ass't Surg. 25th reg't.	Resigned.
Paine, Rob't Treat, Jr.	25	New York University	1861	Attended Bellevue and Black. Isl. Hosp.	Ass't Surg. 29th reg't.	
Perry, Frederick H.	30	New York University	1856	In private practice	Ass't Surg. 29th reg't.	
Peters, Joseph A.	21	Buffalo Medical College	1861	Interne at Alms House for 4 months	Ass't Surg. 21st reg't	
Pettier, Pierre D.	26	Buffalo Medical College	1859			
Phillips, James S.	37	New York University	1855			
Phillips, John P.	27	New York Medical College	1850	In private practice	Ass't Surg. 37th reg't.	
Pitts, James	23	Cincinnati Medical College	1849	In practice 12 years		
Potter, William W.	25	Buffalo Medical College	1857	In private practice 4 years	Ass't Surg. 49th reg't.	
Powell, Richard	23	Royal College Surg., Ireland	1861		Ass't Surg. 88th reg't.	
Priestly, John	40	Glasgow, Scotland	1848	In practice 10 years		
Prentice, Fowler	22	Long Island Hospital	1860	Asst. Phys. at Blackw. Isl. for 5 months	Ass't Surg. 30th reg't.	Prom. to Sur. 73d reg.
Quackenboss, E. M.	25	New York University	1859	18 months in Europe attending Hospital		
Rinelle, M. G.	27	St. Louis	1861			
Radginsky, Louis D.	26	New York University	1859	In practice 10 years	Ass't Surg. 36th reg't.	Resigned.
Ramsay, George M.	37	Jefferson Medical College	1852		Ass't Surg. 95th reg't.	
Rappold, Julius C.	25	New York University	1861		Ass't Surg. 52d reg't.	
Reed, James A.	38	College Phys. and Surgeons	1848	2 years Hospital practice in New York	Ass't Surg. 69th reg't.	
Regan, Matthew F.	27	Buffalo Medical College	1856	Had charge Brooklyn City Hospital, &c.	Ass't Surg. 28th reg't.	Discharged.
Robinson, Joseph W.	32	Jefferson Medical College	1860	In private practice	Ass't Surg. 82d reg't.	
Rogers, J. H.	26	New York University	1859			
Root, Henry	27	College Phys. and Surgeons	1860		Ass't Surg. 54th reg't.	
Ruggles, Augustus D.	39	Albany Medical College	1855		Ass't Surg. 63d reg't.	
Rulison, William H.	30	Albany Medical College			Ass't Surg. 15th reg't.	
Sattler, Cornelius	30	Wurberg				
Schwarzenberg, George	30	Giessen	1852			
Schoon, James H.	36	Albany Medical College	1849	In private practice		

ASSISTANT SURGEONS.—Continued.

NAME.	Age.	Where graduated.	Year.	What service since graduation.	Where appointed.	What changes.
Sweeney, James	25	Albany Medical College	1860	Albany City Hospital 1 term	Captain 96th regiment	
Severin, A.	30	Hamburg	....	2 years practice in Hamburg; 5 in N. Y.		Dismissed.
Shanahan, David Reid	39	College Surgeons, England	1846	Attended St. Vincent's Hospital	Surg. 63d reg't	Prom. to Sur. 87th reg.
Skilton, Julius A.	28	Albany Medical College	1855	In private practice	Ass't Surg. 30th reg't	
Smith, J. E.	30	College Phys. and Surgeons	1854	2 years charge Wayne Co. Hospital		
Smith, Joseph H.	39	Albany Medical College	1846			
Spaulding, Elbridge G.	26	College Phys. and Surgeons	1860		Ass't Surg. 94th reg't	
Spencer, John	38	Cleveland Medical College	1842	Marine Hospital at Cleveland 3 years	Surgeon 9th cavalry	
Sprague, H. M.	25	New York University	1860	One year New York City Hospital		
Stolger, Joseph	28	Wurzburg	....	Hospital practice Vienna and Prague, 1854–1859		
Stein, Charles	38	Eslangen	....		Ass't Surg. 58th reg't	
Steinert, George	29	New York University	....		Ass't Surg. 77th reg't	
Stevens, George T.	28	Castleton, Vermont	1857			
Streeter, Buel G.	29	Castleton, Vermont	1853	In private practice	Ass't Surg. 47th reg't	
Tanner, William H.	24	Vermont Medical College	1860		Ass't Surg. 41st reg't	
Thomain, Robert	36	Switzerland	....	Practiced in Hospital of Paris		
Tissot, Max	29	Bern, Switzerland	1860			
Todd, George B.	30	Albany Medical College	1856	1½ years at Mobile Hospital, Alabama	Ass't Surg. 12th reg't	
Tompkins, Hartwell C.	33	Woodstock	1853		Ass't Surg. 61st reg't	
Town, Francis L.	25	Hanover School, N. H.	1859	8 mo's Island Hospital; 6 mo's New York Hospital	Ass't Surg. U. S. A.	
Trenor, John, Jr.	30	College Phys. and Surgeons	1856		Ass't Surg. 80th reg't	Resigned.
Tuthill, Robert K.	28	New York Medical College	1859		Ass't Surg. 59th reg't	
Uhline, Stephen P.	37	Castleton, Vermont	1851			
Ulter, A.	42	College Phys. and Surgeons	1847	In active practice		
Valentine, S. B.	29	Albany Medical College	1855	Assistant at Albany Hospital one year	Ass't Surg. 3d reg't	
Van Rensselaer, J. J.	25	Albany Medical College	1859	In private practice	Ass't Surg. 1st reg't	
Van Steenberg, Wm.	29	University of Vermont	1856	In private practice		
Van Vleck, D. P.	40	Albany Medical College	1842			
Van Vorst, G. W.	24	College Phys. and Surgeons	1861			
Vaughan, C. H.	30	Albany Medical College	1855	In practice 6 years	Ass't Surg. 96th reg't	
Vosburgh, Benj. F.	25	Albany Medical College	1858	In practice 3 years		
Wallace, Theodore C.	32	Geneva Medical College	1850	In practice 7 years; ship surgeon 2 years	Ass't Surg. 93d reg't	
Welch, Abraham	33	University Pennsylvania	1853			

Werner, Edward	36	Wurzburg	1850	In private practice	....
West, Joseph E.	33	College Phys. and Surgeons	1852	Charge of hospital 18 months at Utica	Ass't Surg. 14th reg't.
Whiton, H. B.	34	Albany Medical College	1854	In private practice	Ass't Surg. 2d reg't.
Whitehead, Ira C.	27	Berkshire Medical College	1855	....	....
Whitford, Alfred H.	33	University Maryland	1856	In private practice 6 years	Ass't Surg. 99th reg't, "Union Coast G'rd."
Williams, Alfred A. C.	26	Berkshire Medical College	1857	Assist. Surgeon 1st Maine Militia 3 months Regiment; ship surgeon	Ass't Sur. 1st reg. artil' Prisoner.
Wisen, William H.	51	Fairfield Medical College	1831	....	Ass't Sur. 2d reg. artil.

Several of the Regiments were independent in their organizations, and were at once accepted for service by the General Government. The Surgeons were appointed by their Colonels, without an examination before the Medical Board. Already in the service of the government it was not easy to recall them for the examination, and they were subsequently commissioned by the State, as follows:

SURGEONS.

NAME.	Age.	Where graduated.	Year.	What service since graduation.	Where appointed.	What changes.
Barr, George W					Surgeon 64th regiment	
Bostwick, Henry P					Surgeon 73d regiment	Resigned.
Elliott, Frederick					Surgeon 1st Cavalry	
Gray, Charles					Surgeon 11th regiment.	
Hindman, Richard H.					Surgeon 67th regiment.	
Lewis, William C.					Surgeon 82d regiment.	
Loughran, Robert.					Surgeon 80th regiment.	
McDonald, James E.					Surgeon 79th regiment.	
Osborne, John Q.					Surgeon 42d regiment.	
Petard, Felix.					Surgeon 55th regiment.	
Powell, Alfred.					Surgeon 82d regiment.	Prisoner at Bull Run.
Simpson, George B. F.					Surgeon 62d regiment.	

ASSISTANT SURGEONS.

NAME.	Age.	Where graduated.	Year.	What service since graduation.	Where appointed.	What changes.
Adams, George					Assistant Surgeon 67th regiment	
Arthard, Theodore					Assistant Surgeon 55th regiment	
Ash, James					Assistant Surgeon 71st regiment	
Calhoun, James T.					Assistant Surgeon 74th regiment	
Fitch, James E.					Assistant Surgeon 79th regiment	Ellsworth Fire Zouaves. [role.]
Forshee, John M.					Assistant Surgeon 11th regiment.	Prisoner at Bull Run. (On parole.)
Furgeson, James F.					Assistant Surgeon 82d regiment	Prisoner at Bull Run. (Parole.)
McLetchie, Andrew					Assistant Surgeon 79th regiment	
O'McDonald, Whoolan					Assistant Surgeon 65th regiment	"Garibaldi Guard."
Ribback, Rudolph					Assistant Surgeon 39th regiment	
Ridgway, Frank					Assistant Surgeon 73d regiment	

As these pages are going through the press I gain the opportunity to add, that when in the month of April a great battle was anticipated at Yorktown, Virginia, where General McClellan was besieging the Confederate army, it was found that the medical and surgical force of the army was insufficient to meet the demands that such an engagement was likely to impose upon it. In order, therefore, to give immediate care to the wounded, under the authority of the Secretary of War, the governors of the several loyal States were directed to appoint a CORPS OF VOLUNTEER SURGEONS, who should respond to the call of the Governor and serve without remuneration. The following appointments have been made by the Governor of the State of New York up to this date, June 11, 1862.

LIST OF VOLUNTEER SURGEONS APPOINTED.

Name.	Residence.	Date of Commission.
James R. Wood	New York city	April 7
Alfred C. Post	do	do
Ernest Krackowizer	do	do
Stephen Smith	do	do
Charles D. Smith	do	do
George A. Peters	do	do
John O. Stone	do	do
Thaddeus M. Halstead	do	do
Willard Parker	do	do
Gurdon Buck	do	do
Lothar Voss	do	do
Thomas M. Markoe	do	do
Alden March	Albany	do
John Swinburne	do	do
Edward H. Parker	Poughkeepsie	do
Charles Winne	Buffalo	do
William Detmold	New York city	do
Mason F. Cogswell	Albany	April 16
Samuel G. Wolcott	Utica	do
Sanford B. Hunt	Buffalo	do
Lewis Post	Lodi, Seneca Co.	do
Jonathan Kneeland	South Onondaga	April 17
John J. Crane	New York city	do
George Cochrane	Brooklyn	do
E. W. Alba	Angelica, Allegany Co.	April 16
Gilson A. Dayton	Mexico, Oswego Co.	April 17
S. Oakley Vanderpool	Albany	do
Daniel E. Kissam	Brooklyn	April 17

5

Names.	Residence.	Date of Commission.
Cornelius Olcott	Brooklyn	April 17
Daniel Ayres	do	do
David L. Rogers	New York city	do
William H. Thompson	do	do
Charles Skinner	Malone, Franklin Co.	April 19
F. Burdick	Johnstown, Fulton Co.	May 15
Smith Ely	Newburgh	do
James V. Kendall	Baldwinsville	May 16
John V. Lansing	Albany	May 19
Sylvester D. Willard	do	do
William S. Denniston	Newburgh	June 9
Benjamin E. Bushnell	Little Falls	June 7
W. Blaisdell	Coeymans	12
Charles H. Porter	Albany	

FORTRESS MONROE AND WHITE HOUSE HOSPITAL.

Yorktown was evacuated on the 4th of May, and the battle at Williamsburg was fought on the 5th and 6th instant.

Dr. John Swinburne of this city, and myself left under the direction of the Surgeon General for Fortress Monroe, on the 8th inst., with orders to report to the medical director, Dr. J. M. Cuyler. We rounded Old Point Comfort after a delightful sail down the Chesapeake, on Saturday morning the 10th inst. Here everything began to have the aspects of war. A large number of vessels of war and transports were lying off the Point. The guns from every point frowned from the massive walls of the Fortress. The rebel flag could be seen at Sewall's Point, and beyond nearly as far as the eye could reach lay the terror to our fleet, the iron clad Merrimac. The Monitor, its antagonist, lay at a little distance from the landing, and would scarcely have gained attention except as she was pointed out to us. On reaching the shore we were escorted to the office of the Provost Marshal, where, although we were both opposed to swearing generally, we swore allegiance to our country with unquestionable earnestness.

Immediately we went to report to Dr. Cuyler, whom we found in a ward of the hospital. He gave us a cordial reception, saying: "You are just the men we want to see, take off your coats and go directly to work here." A large room full of wounded men were before us. Three hundred had arrived the evening previous, by boat, from Williamsburg, where they were wounded on the day of the battle, and had received only field dressing. They had been brought a mile or two to the boat, some of them having lain on the field over night, then removed from the boat, and brought to the hospital. Thus, it was four days since some of the wounds had been dressed, and the patients were suffering from exhaustion As soon as we could procure sponges, basins, water, lint, bandages, straps, &c., we went to work, and hard work it was to bend for hours over the beds of those poor fellows. There were many cases of field amputations, and most of them were good operations. There was, however, a tendency to gangrene, and some of the wounds became fatally gangrenous. There were gunshot wounds of almost every variety, a record of which would be interesting, but there was no time for making it. Many of the patients in the ward assigned to Dr. Swinburne and myself were of the Fifth North Carolina and a Virginia regiment, together with Massachusetts, New York and Michigan regiments. In the wards of the hospital they lay side by side, as amiable towards each other as if they had never been combatants, receiving alike all that care and skill could bestow. There were in our ward seven fractured femurs, occasioned by bullets. The minie ball shatters the femur fearfully, breaking it into splinters of from one to six inches in length, the results of which are likely to prove fatal. These cases ought to be brought off from the field and to the hospital on stretchers, with simple extension to keep the limb straight. Felt splints and bandages, when applied, become tightened by the swelling of the limb, and when the dressings are deferred for two or three days, they are exceedingly painful and productive of great mischief. We had arrived at the hospital at a favorable time for hard work, and we strove to perform it in behalf of our patients, in justice to our profession, and to the State of New York, which we represented. The surgeon in charge of the general hospital was our friend Dr. Reed B. Bontecou, Brigade Surgeon in General Wool's Division, formerly of Troy, to whom we were indebted for hospitality and for many kind attentions. The hospital was a part of the Hygeia Hotel, formerly a fashionable Southern resort, a sort of Southern Saratoga, built to accommodate twelve hundred guests, where the gay and the happy resorted, to breathe the invigorating air from the ocean. It was a sad thought, that where only cheerful voices once mingled, the groans of the wounded and the dying now burdened the air. The roses in the court-yard which once emitted sweet perfume, now seemed sickly, unattractive, and exhaling only the odor of pus and suppurating wounds. There were said to be at the Hygeia about five hundred patients. It was the nearest hospital to the landing, and consequently many of the worst cases were taken off here. After working almost without intermission on Saturday, Sunday and Monday, we found two platform car loads of wounded just arrived. They were laid on the piazza and in the yard, in the rear of the hotel, where their wounds were dressed. Every place seemed covered. They were immediately sent on a transport North.

The capital operations that occurred during our stay were four cases of resection of the shoulder joint, three amputations of the thigh, one amputation by disarticulation at the knee joint, two cases of resection at the elbow, and ten cases of exsection of the femur, eight of which were performed by Dr. Bontecou, and two by Dr. Swinburne. Dr. Bontecou is a graceful and accomplished operator, and must be ranked among the first American surgeons. The result of many of the cases of exsection of the femur were unsuccessful, though not under circumstances that ought to weigh against the operation entirely. The muscles in these cases were greatly torn, and destroyed by the force of the falls; the patients were already exhausted, and the air of the over-crowded wards had become so pus poisoned, that a well man could scarcely have lived a week in them, and it necessitated the vacating of some of them entirely. This was the case of the Reading Room ward of which we had the charge. There were at this hospital, as assistants to Dr. Bontecou, Dr. Van Steenberg, of the First regiment, and Dr. Forshee, of the Eleventh regiment; Brigade Surgeon Shipman, and Dr. Light. One mile west was the Mill Creek Hospital, a large government storehouse, with about three hundred beds, all of which were occupied, under the charge of Brigade Surgeon Hunt. Here Dr. McLean, of the New York Second, and Dr. Whiton, assistant, were stationed. Drs. Brinsmade, of Troy; Alden and Henry March; Lente,* of Cold Spring, and others, were doing volunteer service at this hospital. One mile still further west was the Chesapeake Hospital, a large building formerly known as the Chesapeake Female Seminary. It was under the charge of Dr. McCay, Brigade Surgeon. Drs. Edward H. Parker, Stephen Smith, Husted, and A. C. Post were on service there. I saw Dr. Post apply a ligature to the primitive carotid. He mentions the operation in a letter to the American Medical Times, of June 7. In this

* See description of Dr. Lente, Am. Med. Times, June 14.

establishment there were said to be seven hundred beds, makin fifteen hundred wounded and sick at Old Point Comfort. There were many sad, sad scenes in the hospitals. Among the faithful laborers at the Hygeia was the Chaplain of the United States ship Chesapeake, and Mr. Barcley, a christian philanthropist from Philadelphia, who was unremitting in his attentions to the sick, procuring for them all that money could purchase, and encouraging them by words and acts of kindness.

Drs. Cogswell and Lansing of Albany arrived at Fortress Monroe on Saturday the 17th, when we all received orders from Dr. Cuyler to report to Dr. Tripler, the medical director of the Army of the Potomac, at its headquarters. We reached Yorktown on Saturday evening, and for want of a pilot remained there until morning, having time to make a hurried survey of the place, but not to visit the hospitals, which were under the charge of Dr. Greenleaf. There were fifteen hundred sick in the hospitals at Yorktown.

It was a delightful sail up the Pamunkey river. There was a stillness becoming the Sabbath morning, the clouds were so beautifully reflected in the river that one might question whether he was sailing through the sky or the water. On the banks of the river all was quiet, and except in few places where " contrabands " gathered about their cabins, they appeared deserted. At West Point we had taken on board a "secesh" pilot, without any special guarantee that our steamer would be safe in his hands. We passed safely the vessels that had been sunk in the river to obstruct navigation, and a little past meridian approached Cumberland, where we found the rear of the army. White House, which is the head of navigation on the Pamunkey river, was several miles beyond, and to this place the river was literally crowded with steamers and transports of every description. It is estimated that there could not have been less than ten thousand vessels, steamers and transports. General McClellan and the advance of the army were at this place. One can only be impressed with the magnitude of an army by actually seeing it and being in its commotion. We found our way to the medical director's whose tent was near General McClellan's, and presented our credentials, with the assurance that we were ready for any service. After a little hesitation we were informed that he had nothing for us to do, and the order for our transportation to

Fortress Monroe was accordingly furnished. We spent the afternoon on the field, meeting at almost every point some familiar face. The evening dress parade of the army excited our admiration as the air echoed with the music of a hundred bands. As far as the eye could reach the field was covered with men, and tents, horses, mules, and army wagons. Towards evening the army received orders to move forward at 4 o'clock the next (Monday, May 19,) morning. When we arrived at the steamer in the evening, preparatory to our return, we found a message from Dr. Tripler, requesting us to report to him the next morning at seven o'clock. This we did. He informed us that he had determined to organize a field hospital at that place, and to send back the sick and disabled of the army there for treatment. He requested us to establish this hospital, of which Brigade Surgeon J. H. Baxter was to remain as director. About three hundred sick had been left on the ground. The hospital was to be composed of one hundred tents erected in double line on an oblong square, to accommodate twelve hundred patients or twelve in each tent. Two companies from the New York 93d and one from the 106th Pennsylvania were detailed for the labor under our supervision. There was a delay in obtaining spades and axes; nothing could be done without them. It began to rain early in the afternoon, and the sick men were picked from the road side as fast as tents were erected to shelter them, others gathered under the trees until tents were ready. Night came and there was neither straw or any food. These poor sick and tired fellows laid down on the ground like brave men, without straw or food, and without a word of complaint. On Tuesday ambulances arrived with the sick faster than we were able to dispose of them. The straw that we obtained was wet and musty. There was yet no means for getting water, or beef, or kettles, or wood, and the thousand other things that pertained to the necessities of a hospital, and when night came again we all laid down on the ground in our tents, tired and hungry, and full of sympathy for the sufferings we could not relieve. On Wednesday the army supplies began to come in. The sanitary commission arrived and furnished us with beef, straw, beds, pillows, shirts and towels; while camp kettles, medical stores, coffee, rice and sugar were furnished from the army department. An arrangement was made for the transportation of wood and water; system and comfort began to come out of confusion and want. To Dr. Cogswell was assigned the laborious duties of the office, and the superintendence of the hospital records, while to Drs. Swinburne, Lansing and myself, of Albany, Drs. Page and Hall, of Boston, was entrusted the reception and the treatment of the patients.

On Thursday a tremendous rain flooded the ground and some of the tents, so that many of the sick lay in the water. This was bad enough, but the men were brave and uncomplaining. Hay was brought after the rain, to raise them above the wet, and the surgeons waded through mud nearly to the top of their boots to see that the hay was well distributed, and to look after the sick. Immediate measures were then taken to floor the tents with plank, six inches above the ground, and to increase the drains around them. The Sanitary Commission did excellent service, and provided for the immediate wants of the sick, before the government resources could be obtained. They had the steamers Spaulding, Elm City and Daniel Webster, on which they received about four or five hundred of the most severe cases from the hospital during the first week. There were received at this field hospital during the first week about seventeen hundred patients. Many of them suffered in their re-transportation from the hospital to the steamers, and doubtless the mortality was increased by the removal of exhausted fever patients. When the wagons and ambulances reached the hospital at night, there was no alternative but to leave the patients in them until morning before beds could be provided for them. Perhaps many convalesced by the time they reached New York who would have been ready to join their regiments had they remained. Dr. Kneeland arrived during the week and labored hard and acceptably in hurrying the preparation of provisions for the patients. The culinary department was crude, and needed constant surveillance to make it run well. The patients necessarily suffered for food for the first few days. But the mortality was not large. During the first week there were only four deaths at the hospital, out of the seventeen hundred. Eight occurred on board the sanitary vessels, possibly some of these might have been avoided if their removal could have been prevented. I have not the statistics of sickness of the whole hospital, but those of two hundred and thirty-four patients for which I prescribed in the morning of May 24th, from which my report was furnished, they may be taken nearly as an index to the whole, and are as follows :

Two Hundred and Thirty-four Patients Visited and Prescribed for on the morning of the 24th of May, were as follows.

Debility	94
Fever	45
Diarrhœa	49
Rheumatism	16
Dysentery	6
Lame and wounded	8
Measles	3
Ruptured	3
Parotitis	1
Venereal	1
Pneumonia	1
Eruptive	3
Injured sight	1
Neuralgia	1
Spermatorrhœa	1
Sore throat	1
	234

Many of these. cases would be able to return to duty in ten days. They were tired and exhausted; they needed REST and NOURISHMENT. The worst cases of fever, both remittent and typhoid, were sent to the vessels of the Sanitary Commission. Could they have been as well nourished at the hospital, they would have gained nothing by removal. The diarrhœas were not unusally obstinate, nor the dysentery severe, and but few of the cases of rheumatism were acute. The stimulus and tonic consisted of quinine and whisky, and were essential in the treatment of nearly every case. Suitable nourishment for the sick would have frequently answered better, but that could not be obtained. The resources of the government were large, but it required several days to concentrate them for the care of so many hundred sick. The zeal and energy of the Sanitary Commission, their timely aid at this hospital, the heart with which they came to the work, the willingness of their medical corps, rise above the praise which words can express. Such, in few words, was the organization of the hospital at Whitehouse, twenty-three miles from Richmond, and the part which those who represented the State of New York bore in it. It was a great labor, and faithfully performed.

The situation which Dr. Tripler chose for the location of the hospital, on low spongy ground, has been severely criticised, and the conduct of General McClellan as severely censured for guarding the house of the rebel Colonel Lee (because it was the place where Washington first met his wife, Mrs. Custis,) instead of allowing it to be used for a hospital. I went through the house. It would not have contained fifty beds, and the lawn was not large enough for so many tents. The river was on one side of it and the road on the other. The miasm from the river and the noise from the road would have been two objections against using the lawn. A more desirable location might have been obtained on a slight elevation at a considerable distance south of Colonel Lee's house, where water would have been more easily available from the river, and the ground more easily surface drained, but Colonel Lee's house would have furnished no desirable addition to a hospital unless used for the headquarters of the surgeons or officers who might be sick. In the present case it was too far distant for the former purpose.

On Sunday the 25th of May, a delegation of twenty-four surgeons arrived from Massachusetts, and engaged with Dr. Tripler for service. Twelve were retained at the White House hospital and twelve sent back to Yorktown.

Circumstances obliged Dr. Cogswell and myself, though not without many regrets, to start on our journey homeward on the 26th instant, while Drs. Swinburne and Lansing went on and were at Savage's Station in time to receive hundreds of the wounded at the terrible battle of Fair Oaks, near Richmond, which occurred a few days afterward.

There are many physicians who have recently gone to do volunteer service, and the many who in various ways are connected with the labors of the sanitary commission, whose names and an account of whose services cannot at the present time be obtained.

June 12, 1862.